An Apartment House Close Up

An Apartment House Close Up

by PETER SCHAAF

FOUR WINDS PRESS   NEW YORK

Library of Congress Cataloging in Publication Data
Schaaf, Peter.
An apartment house close up.

SUMMARY: A photographic introduction to all the parts of an apartment house: doors, windows, elevators, hallways, rooms, laundry, etc.

1. Apartment houses — Pictorial works — Juvenile literature. [1. Vocabulary. 2. Apartment house — Pictorial works] I. Title.
TH4820.S3   690'.8314   80-11301
ISBN 0-590-07670-1

PUBLISHED BY FOUR WINDS PRESS
A DIVISION OF SCHOLASTIC MAGAZINES, INC., NEW YORK, N.Y.
COPYRIGHT © 1980 BY PETER SCHAAF
ALL RIGHTS RESERVED
PRINTED IN THE UNITED STATES OF AMERICA
LIBRARY OF CONGRESS CATALOG CARD NUMBER: 80-11301
1 2 3 4 5 84 83 82 81 80

for Brooke

DOORS

HALLWAYS

ELEVATORS

ROOMS

WINDOWS

HEAT

COLD WATER

HOT WATER

GAS

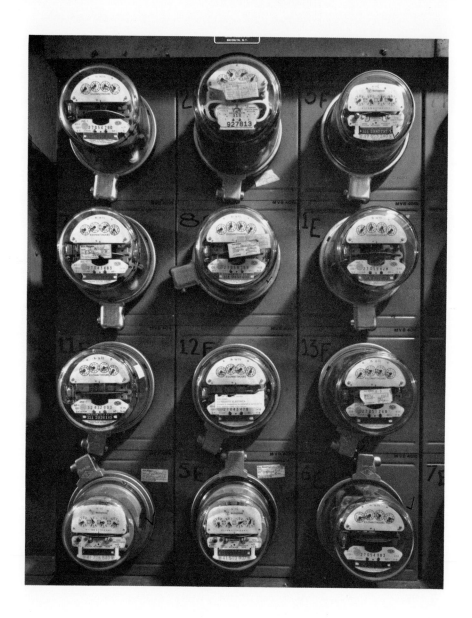

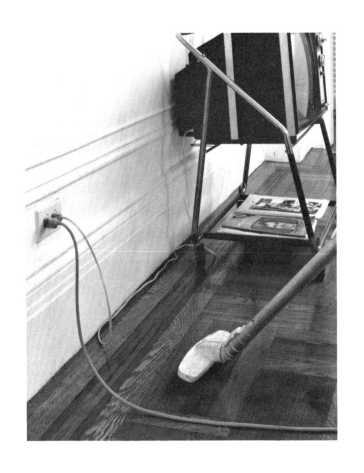

ELECTRICITY

GARBAGE

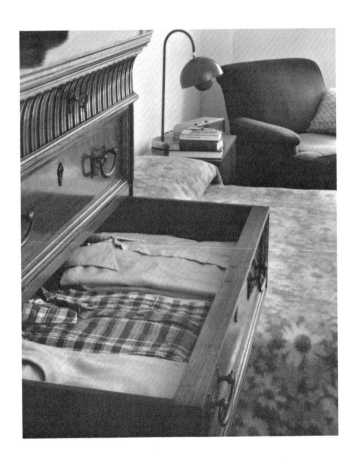

HOME